Texts and Emails are Ruining My Life!

Laurie Leiker

Copyright © 2015 Laurie Leiker

All rights reserved.
ISBN: 1518724221
ISBN-13: 978-1518724220

DEDICATION

For Sarah & Kenny

CONTENTS

	Acknowledgments	i
1	Am I Addicted?	1
2	Email/Text Truths	6
3	Texting and Email Do's	10
4	Email and Texting Do's	20
5	Wrapping It Up	28

ACKNOWLEDGMENTS

No one writes a book in a vacuum and this one is no exception. I couldn't have done this without the support of my family, Sarah, Kenny, Cherri, Sara and Jen; they put up with my intensity during the writing process. I also need to thank my editors: Gregory Martin, Jennifer Cleaveland and Shelley Delayne Siracusa.

Photo by: Taco Ekkel/Creative Commons

1 AM I ADDICTED?

Email. Texting. Instant communications. In the past, we wanted a way to be able to reach out to anyone whenever we want. Now, we can't get away.

Email has become the new monster under the bed, the new pile of letters in the middle of the dining room table. Trust me – there's no difference between that pile on the dining room table and the 300+ emails scattered throughout your email service, whether you use a free public email address like Gmail, Outlook or Yahoo or you have a separate email service.

Texting has raised the bar even higher. So many of us communicate only through text because it's easier, faster and we can have an immediate answer.

The fact is, we're addicted to non-personal communications – both email and, more often, texting. We check all the time, day and night, to make sure we haven't missed anything. We're disappointed when there is nothing new. We save email and texts "just in case."

Email and text addiction – it's a real thing - follows us everywhere – home, office, hobby, sport – you name it, you've got an email or a text about it.

Really, these addictions aren't any different than running to the mailbox as soon as you see the mailman leave. We deal with email and texts the same way we deal with snail mail, only more immediately and with more disorganization. If your dining room table or desk is piled with unopened letters, bills and junk mail, chances are your email inboxes are too.

But how can you tell if you're truly addicted to digital communications? Here are some sure signs:

- Each email inbox contains more than 50 emails all the time.
- Your "sent" folder has more than 1000 emails.
- If you don't get an email or text every 10 minutes, you start to worry there's something wrong.

- If you don't get a response to a text you sent within five minutes, you fire off several other texts just to make sure they got the first one and to chastise them for ignoring you.
- You regularly check your phone while you're in the bathroom.
- Your friends/family/co-workers complain because you're staring at your phone while they're trying to talk to you.

If you're bothered by any or all of these symptoms (and trust me, it's embarrassing when you're caught checking your phone in the bathroom), you may be an addict.

So, how do you get it under control? Is there a way to stop yourself and truly get it organized so it's not ruling your life? Absolutely!

Photo by: Sanja Gjenero/FreeImages

2 EMAIL AND TEXTING TRUTHS

It seems as if digital communications have always been there, doesn't it? But for a lot of people, it's all they've ever known; heaven forbid you have to wait on the postal service to bring you important information, even if it's just this week's sales at the local grocery store.

Believe it or not, the world actually did function well before email and texting. It thrived and grew over hundreds of years with everything from hand delivery and smoke signals to public post, pony express, mail trains, mail trucks and finally, email. So email and texts are just the latest in the chain of how we communicate.

Email and texts are actually only tools to help us be more productive and keep in touch with others. We have word processors, copiers, printers, spreadsheets, and a few things you've probably never heard of before: the telephone (that black thing hanging on the wall), pen and paper. Yes, these have driven our lives for ages; some even had to work with just the pen and paper before the telephone!

Like a carpenter has many items in his toolbox - a hammer, nails, a saw - e-mail and texts are just items in our communications toolbox. They let us communicate and generally pass information along more efficiently, when it's used correctly.

By the same token, using e-mail and texting means you have to maintain them. In order to actually use his tools, a carpenter has to keep his saw oiled and tuned; so staying on top of your e-mail inbox, deleted items, send items and general set up, as well as how many text messages you are storing on your phone, will help you be more efficient and effective.

A friend called me a few years ago because her email seemed really slow. In fact, one night, it just stopped. She could never find anything in her inbox and she always felt she was drowning in email. I walked her through the steps I'm going to show you in this book three times and she never seemed to be able to get it under control. By the time she called me the fourth time, she had over

3000 items in her inbox, 6000 in her deleted folder and a similar amount in her sent folder. Why? Because each time we talked earlier, she felt it was too much hassle, that she knew a better way and ignored my tips suggestions. After the fourth conversation, she knuckled down, went back through her email and in three hours, was down to nothing in her deleted folder, nothing in her sent folder and 6 emails in her inbox.

Another friend was having a problem with her phone; it was dragging, the battery was draining too quickly and she just couldn't understand why. Turns out she never deleted her old text messages; she said she didn't know if she should. When we went through and deleted everything more than three months' old, then restarted her phone, guess what? Bingo! No more problems with the battery or the phone going slowly.

Getting your texts and emails under control is only one problem, and a minor one at that. Almost hourly, someone is getting upset by an email or text that's been misunderstood because the sender said something in a joking manner, but the joke didn't come through the way it was intended.

Has that happened to you? To help you avoid losing friends and colleagues, let's look at some texting and email Dos and Don'ts.

Photo by: Slvesh Kumar/Creative Commons

3 TEXTING AND EMAIL DOS AND DON'TS

How many emails and texts do you get a day? Dozens, no doubt, most of which are trash. If you're on social media, have your work email coming into your home email inbox, or just have a lot going on, the number of emails and text messages you receive can be upward of 100 each day.

Sound unreasonable? Maybe, but we all do it – with instant communications, we make full use of it and expect everyone else is sitting at their desks or on their phones, waiting to pounce on the first email or text that jumps into their line of sight. There's nothing worse than sending a text, expecting an instant answer and having to wait even 10 minutes for an answer. It's instant communications, after all!

It's easy to forget others have personal lives and don't always sit by their phones, just in case someone texts or emails them. I know – hard to believe, but several people very close to me have told me it is true.

The worst thing you could do when someone doesn't reply to your texts or emails right away is to set off a storm - sending an email or text, then sending another, then another, all in rapid-fire succession until the person you're sending to gets upset and/or answers. In the end, if the you just take a breather for a second, set your phone down and walk away, get something to drink or actually talk to someone, you could find a way to relax about your electronic communications and become less annoying, less bullying and make more sense. It does feel that way sometimes, doesn't it? That incessant ping-ping-ping-ping of rapid-fire emails and texts? It's like someone poking you in the forehead over and over – you just want to smack them.

So what can you do? To start, don't be part of the problem. Stop and think before you send an email or text:

- Are you angry? STOP. Don't send anything until you've calmed down and have a bit more perspective.

- Is this necessary? Do you absolutely HAVE to send this?
- Is it useful? What information are you trying to convey?
- Does everyone you're sending it to need it?
- Can it be done by calling?
- Can it be done smarter?

Let's look at a scenario:

Mom was angry when she found out her daughter, away at college, charged a large amount on her credit card when she was home a few weeks before. Mom sent a text. When Daughter didn't respond right away, Mom sent another text. Then another. A total of 10 messages within the space of 15 minutes. The result? Daughter apparently was in class and had to turn her phone off; she didn't see any of the texts until she turned her phone on again and was understandably upset to see the firestorm taking place on her phone.

Ok. Take a breath and let's look at it in light of our 6 questions. Was Mom angry? Obviously, which means she should have put her phone down BEFORE sending the first text.

Was it necessary? Probably; she needed to talk to her child.

Was it useful? Not really. She was angry, so she didn't convey any information other than the fact that she was angry and considered her child worthless, so it accomplished nothing except alienating the child.

Could it have been taken care of on the phone? Absolutely!

Could it have been done smarter? Obviously.

Before you blast a series of texts or emails, remember that the person you're sending to is probably busy; they might be in the bathroom, taking a shower, in a meeting or class or somewhere else a smartphone would be inappropriate. It might take a few minutes or a bit longer to get a response. If you need an immediate response, pick up the phone.

If you get their voice mail, you'll know the other person is busy and you can leave a voice message.

Email and Texts Are One-Dimensional

Ever get one of those texts that seem just rude and crude? What did you think of the sender? Not much, probably. But in reality, the sender was probably just joking, assuming you knew what he/she was really saying.

Thus comes the dilemma with electronic communications – they're one-dimensional. The reader can't see your body language, can't see if you're smiling or not, probably doesn't know your sense of humor. The twinkle in your eye when you're joking doesn't come through no matter how many emojis you use.

We all have a tendency read electronic communications based on the way we're feeling at the time. What does that mean? When we read, the voice we hear is ours and how we're feeling at that moment is the tone with which the email is read. If you're feeling down, you're more likely to take any jokes or innuendos as insults, regardless of how they're meant.

So how do you avoid problems? It's really not that complicated, with just a few steps:

- Simply read your message before sending, especially if your message is urgent or if you're sending something to a business colleague or client. Check for typos or places where auto correct has gone crazy; this is especially important with text messages, which are typically sent more quickly.
- Talk to people. Talk and talk and talk; if you simply rely on email or texts to communicate, you're missing 90% of communications. The more you talk to people, the more they will be able to hear your voice reading the message to them, not their own.

- Really get to know those around you; the more they know you, the more likely they'll understand your meaning, no matter how you're communicating with them.

Email and Texts Are Not Substitutes For Good Communications

A common mistake often made with e-mail and texting is to hide behind it instead of dealing on a personal basis with others.

Take, for instance, Gina, a woman who only communicated with her extended family through text and email. She was confused because she didn't hear much from her family and she wasn't getting responses. We talked about it several times. Finally, when she texted her aunt at lunch one day, the problem became clear; she had no personal communications with her family.

It's important that we use all forms of available communication. For some, emails or texts aren't the best way to communicate with them; for others, it's the only way they know what's going on.

How do you know if you're message is getting through? A good rule of thumb is: If there are more than three emails or texts in a chain without a response, pick up the phone; you're obviously not getting your point across. You could be dealing with someone who does better hearing what you're saying, instead of reading it, and everyone appreciates it when someone picks up the phone and calls.

Content Matters

Email and texting are the most common forms of electronic communication, so content definitely matters.

Whether you're sending an email or text, make your content to the point. Use as few sentences as possible. Use good grammar, spelling and punctuation; remember – while the brain doesn't see each individual letter in a word, if one is missing or incorrectly placed, the eye will stop there and it's difficult to get the reader to continue on; the same goes with punctuation and grammar.

With text messages, double check your message before you hit "send." Please. Most miscommunications and embarrassment through texts happen because auto correct went haywire and something ridiculous was sent.

With emails, remember that most people do not scroll down the screen – not on emails, not on web pages. By keeping your email to the point, there's a better chance of the reader getting to the end.

LAURIE LEIKER

Photo by: Paxon Woelber/Creative Commons

4 EMAIL BALANCING ACT

Getting your electronic communications under control can be a delicate balancing act; it's like trying to eat an elephant – you don't always know where to start, but with one bite at a time, soon, you've reached your goal. With email especially, it takes four bites - assessment, rebalancing, cleaning up and keeping up.

Assessing your email

With any bad habit, it's important to acknowledge you have a problem. How many messages are in your inbox right now? If it's more than 50, you're not just holding things you might need - you're addicted.

What about your deleted box and sent folders in your email? These are two areas most often ignored, but they often are the biggest cause of email slowdown. Your deleted and sent folders are not there for "safekeeping;" if you're saving things in your deleted and sent folders "just in case," chances are (1) you'll probably never need them again; and (2) you've already found them somewhere else.

By keeping your deleted and sent folders cleaned out, you'll automatically be more efficient, find things faster and your system won't bog down as often.

Rebalancing

After looking at how many emails you have sitting around, you need to rebalance. Keep your email (and, of course, your texts) in balance with your other tools. Remember - sometimes, it's easier to pick up the phone than to send an email or a text, especially if you're passing along important information that shouldn't be misunderstood.

The best way to rebalance is to categorize each item as you're touching it. Ask yourself:

- Do I need it?
- Do I want it?
- Does it make me money?

If it doesn't fit into one of those three categories, toss it. You can apply this to almost anything you're cleaning out, not just your email – cleaning closets, organizing your pantry – any area of life.

Cleaning Up

Once you've figured out where the problem is, and it's time to clean up.

Clean up can actually be fun, believe it or not. The more often you do it, the more you'll realize it wasn't such a hard thing to do after all. And you'll immediately feel a sense of relief.

For text messages, it's easy – every phone carrier and model of phone has a way to delete all or some of your older text messages. My recommendation is that you delete text messages older than 10 days; if you haven't dealt with the information by then, it wasn't that important to you.

For email, the first thing to do is to make sure you have folders set up. To set up a folder, right click on your inbox, or go to Folders and click on "new folder." Type in the name of the folder you'd like to set up, hit "enter" and it'll appear.

Once your folders are set up, you can sort your email by subject.

Within your email client, click on "arrange by" or "From" and select "subject;" this groups all your emails by subject so you can see entire conversations. From there, drag and drop the entire conversation into folders, delete older parts of the conversation and save only the last email, or forward the conversation to someone else.

To drag and drop a conversation into a folder, highlight the emails you want to move using shift+down arrow. Once highlighted, drag the group into the right folder or click the "delete" key to delete the whole group.

▲ interview for cancer article	
✉ Liz Seegert Re: interview for cancer article	5/11/2012
✉ Liz Seegert Re: interview for cancer article	5/11/2012
✉ Liz Seegert interview for cancer article	5/11/2012

It helps to sort your email in several different ways to make sure you're getting everything you want into the correct files. Sorting by "from," "attachments" and by "type" will make sure you're not missing anything that needs to be filed or deleted. "Type" is especially helpful when sorting through your sent email; it will group things together for easy deletion.

Moving Forward

The next tip is one of the hardest to get used to, but once you do, it'll help knock down the amount of email and text messages you're hanging onto: Handle each only once, if possible.
- Is it an action item? If so, get it done. NOW.
- Do you need it or can it be found somewhere else? If someone sent you an attachment or a phone number/address, they still have it, so if you don't absolutely need to keep it, delete it!
- Do you want it? File it.

There are reasons to keep things, but you don't have to keep them in your inbox or on your phone – set up files for them!

Now, repeat the same actions with all your electronic communications, including your inbox, your deleted email folder and your sent email folder. When looking at your deleted folder, remember: those emails were deleted for a reason; if you don't empty your deleted folder because you're afraid you might throw out something you need, file it the first time. By now, you should be down to almost nothing in any of your active folders.

Keeping Up

Whew! That might have taken you some time, but it definitely was worth it! Now, it's time to set some things in place to make sure your electronic communications don't return to a jumbled mess.

1. Handle each email and text only once using the list – is this something that must be addressed today? If so, get it done. Is this something that can be done when time permits? File it.
2. For emails, use drag-and-drop not only to file things but also to set up new contacts and calendar items. Most email services let you drag-and-drop into contacts and calendars; it makes adding these things so much easier.

There are an unlimited number of ways you can choose to deal with an email message, from moving it to a specific folder to flagging it for followup. We won't go into more detail on setting up folders here, because it's really a personal choice; just make sure to set them up. Just keep your priority emails in your inbox so they're handy.

Got it? Ok – let's wrap things up.

Photo by: Franklekeon/Creative Commons

5 WRAPPING IT UP

As with anything in life, balance is key to keeping control of texts and emails and not letting them rule your life. If you follow the suggestions outlined, I know you'll be able to manage well, be more productive and efficient and feel less stressed.

Some final thoughts:
- Keep only your important emails and texts so they're in front of you.
- When replying to email, use "reply all" sparingly, if at all – just reply to the sender and let them handle

the rest; not everyone needs to know you thanked the sender.

- Always consider your audience when sending an email or text and remember, they're just like you – they don't want more in their inboxes and on their phones.
- Remember the one-dimensional nature of electronic communications and break that barrier through talking, listening and communicating through all your available tools to keep the lines of communications open.
- Be considerate of others' time. As we mentioned before, we're all busy. Don't get upset or assume you're being ignored simply because you haven't received an immediate response. Unless you're dying, nothing is that important. If you're dying, you shouldn't be texting anyway – call 911.

ABOUT THE AUTHOR

Laurie Leiker has spent the last 20+ years helping consumers manage new technology in better ways. She lives in Austin, Texas and enjoys cooking, reading and writing.

www.ingramcontent.com/pod-product-compliance
Lightning Source LLC
Chambersburg PA
CBHW041618180526
45159CB00002BC/908